3-CUBIC METER BIOGAS PLANT

A CONSTRUCTION MANUAL

Published by

VOLUNTEERS IN TECHNICAL ASSISTANCE
3706 Rhode Island Avenue
Mt. Rainier, Maryland 20822 USA

ACKNOWLEDGEMENTS

This book is one of a series of manuals on renewable energy technologies. It is primarily intended for use by people in international development projects. The construction techniques and ideas presented here are, however, useful to anyone seeking to become energy self-sufficient.

Volunteers in Technical Assistance, Inc., wishes to extend sincere appreciation to the following individuals for their contributions:

William R. Breslin, VITA, Mt. Rainier, Maryland
Ram Bux Singh, Gobar Gas Research Station, India
Bertrand R. Saubolle, S.P., VITA, Nepal
Paul Warpeha, VITA, Mt. Rainier, Maryland
Paul Leach, VITA, Morgantown, West Virginia

TABLE OF CONTENTS

3-CUBIC METER BIOGAS PLANT

A CONSTRUCTION MANUAL

I. WHAT IT IS AND HOW IT IS USEFUL

Biofuels are renewable energy sources from living organisms. All biofuels are ultimately derived from plants, which use the sun's energy by converting it to chemical energy through photosynthesis. When organic matter decays, burns, or is eaten, this chemical energy is passed into the rest of the living world. In this sense, therefore, all life forms and their by-products and wastes are storehouses of solar energy ready to be converted into other usable forms of energy.

The kinds and forms of the by-products of the decay of organic matter depend on the conditions under which decay takes place. Decay (or decomposition) can be aerobic (with oxygen) or anaerobic (without oxygen). An example of anaerobic decomposition is the decay of organic matter under water in certain conditions in swamps.

Aerobic decomposition yields such gases as hydrogen and ammonia. Anaerobic decomposition yields primarily methane gas and hydrogen sulfide. Both processes produce a certain amount of heat and both leave a solid residue that is useful for enriching the soil. People can take advantage of the decay processes to provide themselves with fertilizer and fuel. Composting is one way to use the aerobic decay process to produce fertilizer. And a methane digester or generator uses the anaerobic decay process to produce both fertilizer and fuel.

2

One difference between the fertilizers produced by these two methods is the availability of nitrogen. Nitrogen is an element that is essential to plant growth. As valuable as compost is, much of the nitrogen held in the original organic materials is lost to the air in the form of ammonia gas or dissolved in surface runoff in the form of nitrates. The nitrogen is thus not available to the plants.

In anaerobic decomposition the nitrogen is converted to ammonium ions. When the effluent (the solid residue of decomposition) is used as fertilizer, these ions affix themselves readily to soil particles. Thus more nitrogen is available to plants.

The combination of gases produced by anaerobic decomposition is often known as biogas. The principle component of biogas is methane, a colorless and odorless gas that burns very easily. When handled properly, biogas is an excellent fueld for cooking, lighting, and heating.

A biogas digester is the apparatus used to control anaerobic decomposition. In general, it consists of a sealed tank or pit that holds the organic material, and some means to collect the gases that are produced.

Many different shapes and styles of biogas plants have been experimented with: horizontal, vertical, cylindrical, cubic, and dome shaped. One design that has won much popularityu for reliable performance in many different countries is presented here. It is the Indian cylindrical pit design. In 1979 there were 50,000 such plants in use in India alone, 25,000 in Korea, and many more in Japan, the Philippines, Pakistan, Africa, and Latin America. There are two basic parts to the design: a tank that holds the slurry (a mixture of manure and water); and a gas cap or drum on the tank to capture the gas released from the slurry. To get these parts to do their jobs, of course, requires provision for mixing the slurry, piping off the gas, drying the effluent, etc.

In addition to the production of fuel and fertilizer, a digester becomes the receptacle for animal, human, and organic wastes. This removes from the environment possible breeding grounds for rodents, insects, and toxic bacteria, thereby producing a healthier environment in which to live.

II. DECISION FACTORS

Applications:

- Gas can be used for heating, lighting, and cooking.

- Gas can be used to run internal combustion engines with modifications.

- Effluent can be used for fertilizer.

Advantages:

- Simple to build and operate.

- Virtually no maintenance--25-year digester lifespan.

- Design can be enlarged for community needs.

- Continuous feeding.

- Provides a sanitary means for the treatment of organic wastes.

Disadvantages:

- Produces only enough gas for a family of six.

- Depends upon steady source of manure to fuel the digester on a daily basis.

- Methane can be dangerous. Safety precautions should be observed.

CONSIDERATIONS

Construction time and labor resources required to complete this project will vary depending on several factors. The most important consideration is the availability of people interested in doing this project. The project may in many circumstances be a secondary or after-work project. This will of course increase the length of time needed to complete the project. The construction times given here are at best an estimation based on limited field experience.

Skill divisions are given because some aspects of the project require someone with experience in metalworking and/or welding. Make sure adequate facilities are available before construction begins.

The amount of worker-hours needed is as follows:

- Skilled labor - 8 hours
- Unskilled labor - 80 hours
- Welding - 12 hours

Several other considerations are:

- The gas plant will produce 4.3 cubic meters of gas per day on the daily input from eight cattle and six humans.

- The fermentation tank will have to hold approximately 7 cubic meters in a 1.5 X 3.4 meters deep cylinder.

- A gas cap to cover the tank should be 1.4 meters in diameter X 1.5 meters tall.

COST ESTIMATE

$145-800 (U.S., 1979) includes materials and labor.

---

*Cost estimates serve only as a guide and will vary from country to country.

III. MAKING THE DECISION AND FOLLOWING THROUGH

When determining whether a project is worth the time, effort, and expense involved, consider social, cultural, and environmental factors as well as economic ones. What is the purpose of the effort? Who will benefit most? What will the consequences be if the effort is successful? And if it fails?

Having made an informed technology choice, it is important to keep good records. It is helpful from the beginning to keep data on needs, site selection, resource availability, construction progress, labor and materials costs, test findings, etc. The information may prove an important reference if existing plans and methods need to be altered. It can be helpful in pinpointing "what went wrong?" And, of course, it is important to share data with other people.

The technologies presented in this series have been tested carefully, and are actually used in many parts of the world. However, extensive and controlled field tests have not been conducted for many of them, even some of the most common ones. Even though we know that these technologies work well in some situations, it is important to gather specific information on why they perform better in one place than in another.

Well documented models of field activities provide important information for the development worker. It is obviously important for a development worker in Colombia to have the technical design for a plant built and used in Senegal. But it is even more important to have a full narrative about the plant that provides details on materials, labor, design changes, and so forth. This model can provide a useful frame of reference.

A reliable bank of such field information is now growing. It exists to help spread the word about these and other technologies, lessening the dependence of the developing world on expensive and finite energy resources.

A practical record keeping format may be found in Appendix II.

IV. PRECONSTRUCTION CONSIDERATIONS

The design presented here is most useful for temperate or tropical climates. It is a 3-cubic meter plant that requires the equivalent of the daily wastes of six-eight cattle. Other sizes are given for smaller and larger digester designs for comparison.

This digester is a continuous-feed (displacement) digester. Relatively small amounts of slurry (a mixture of manure and water) are added daily so that gas and fertilizer are produced

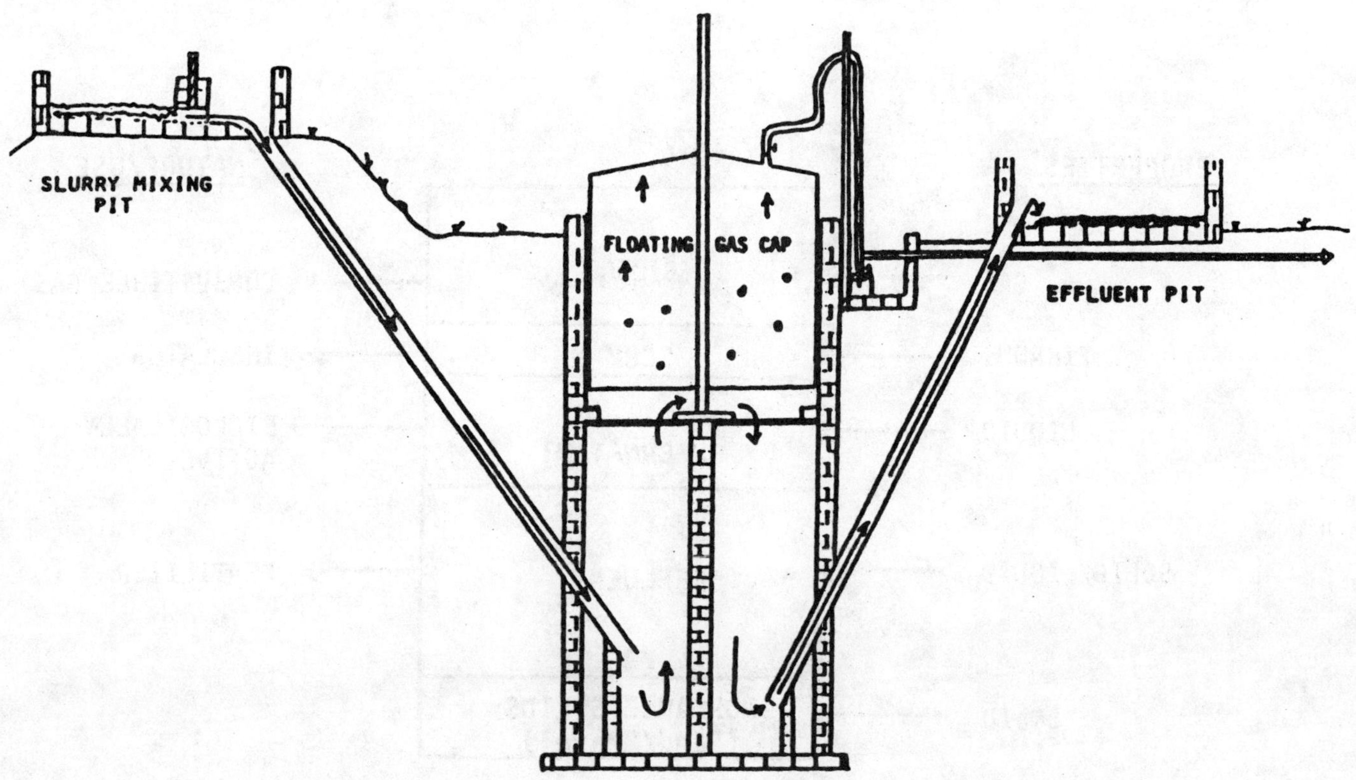

Figure 1.   3-Cubic Meter Biogas Digester

continuously and predictably. The amount of manure fed daily into this digester is determined by the volume of the digester itself, divided over a period of 30-40 days. Thirty days is chosen as the minimum amount of time for sufficient bacterial action to take place to produce biogas and to destroy many of the toxic pathogens found in human wastes.

BY-PRODUCTS OF DIGESTION

Table 1 shows the various stages of decomposition and the forms of the material at each stage. The inorganic solids at the bottom of the tank are rocks, sand, gravel, or other items that will not decompose. The effluent is the semisolid material left after the gases have been separated. The supernatant is biologically active liquid in which bacteria are at work breaking

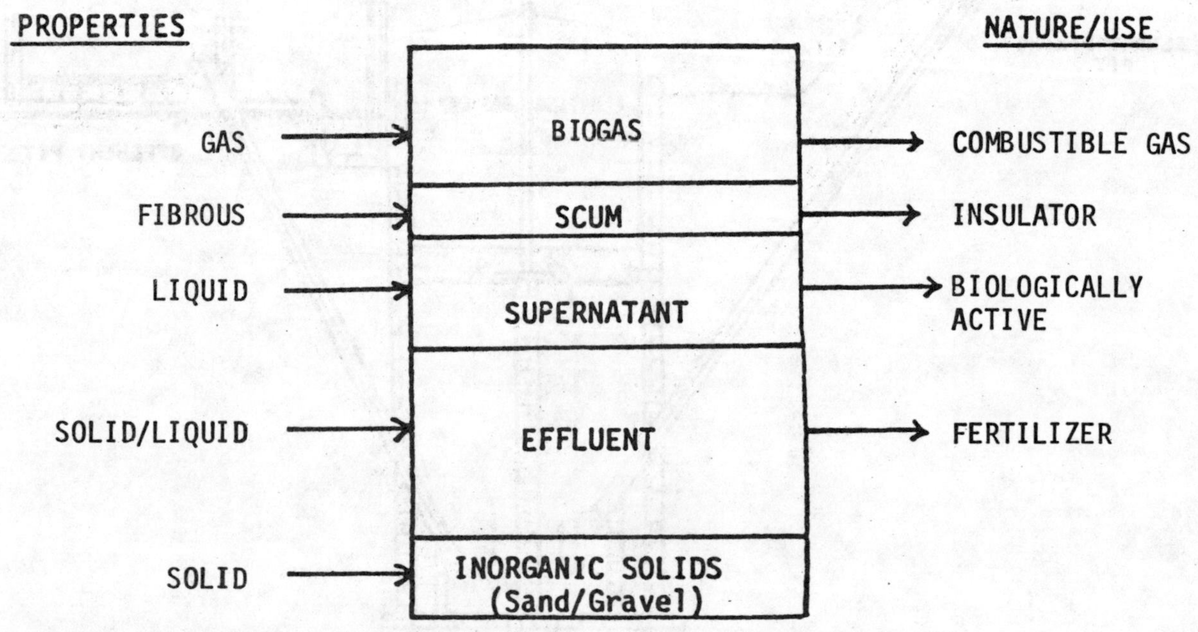

Table 1. Anaerobic Decomposition of Organic Material in Biogas Digesters

down the organic materials. A scum of harder-to-digest fibrous material floats on top of the supernatant. It consists primarily of plant debris. Biogas, a mixture of combustible (burnable) gases, rises to the top of the tank.

The content of biogas varies with the material being decomposed and the environmental conditions involved. When using cattle manure, biogas usually is a mixture of:

CH_4 (Methane)	54-70%
CO_2 (Carbon Dioxide)	27-45%
N_2 (Nitrogen)	.5-3%
H_2 (Hydrogen)	1-10%
CO (Carbon Monoxide)	0-.1%
O_2 (Oxygen)	0-.1%
H_2S (Hydrogen Sulfide)	

Small amounts of trace elements, amines, and sulphur compounds.

The largest, and for fuel purposes the most important, part of biogas is methane. Pure methane is colorless and odorless. Spontaneous ignition of methane occurs when 4-15% of the gas mixes with air having an explosive pressure of between 90 and 104 psi. The explosive pressure shows that biogas is very combustible and must be treated with care like any other kind of gas. Knowledge of this fact is important when planning the design, building, or using of a digester.

LOCATION

There are several points to keep in mind before actual construction of the digester begins. The most important consideration is the location of the digester. Some of the major points in deciding the location are:

. DO NOT dig the digester pit within 13 meters of a well or spring used for drinking water. If the water table is reached

when digging, it will be necessary to cement the inside of
the digester pit. This increases the initial expense of
building the digester, but prevents contamination of the
drinking supply.

. Try to locate the digester near the stable (see Figure 2) so
excessive time is not spent transporting the manure. Remember, the fresher the manure, the more methane is produced as
the final product and the fewer problems with biogas generation will occur. To simplify collection of manure, animals
should be confined.

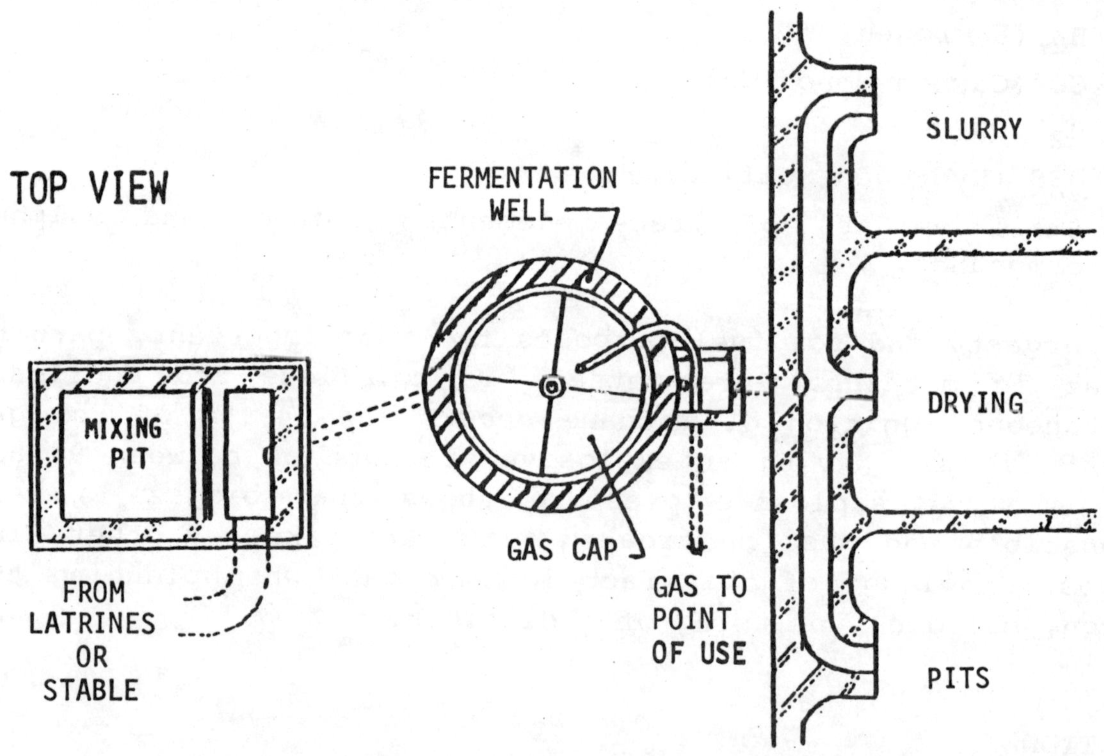

Figure 2. Location of Digester From Fuel Source

. Be sure there is enough space to construct the digester. A
plant that produces 3 cubic meters of methane requires an
area approximately 2 X 3 meters. If a larger plant is
required, figure space needs accordingly.

. Arrange to have water readily available for mixing with the manure.

. Plan for slurry storage. Although the gas plant itself takes up a very small area, the slurry should be stored either as is or dried. The slurry pits should be large and expandable.

. Plan for a site that is open and exposed to the sun. The digester operates best and gives better gas production at high temperatures (35°C or 85-100°F). The digester should receive little or no shade during the day.

. Locate the gas plant as close as possible to the point of gas consumption. This tends to reduce costs and pressure losses in piping the gas. Methane can be stored fairly close to the house as there are few flies or mosquitos or odor associated with gas production.

Thus, the site variables are: away from the drinking water supply, in the sun, close to the source of the manure, close to a source of water, and close to the point where the gas will be used. If you have to choose among these factors, it is most important to keep the plant from contaminating your water supply. Next, as much sun as possible is important for the proper operation of the digester. The other variables are largely a matter of convenience and cost: transporting the manure and the water, piping the gas to the point of use, and so on.

SIZE

The amount of gas produced depends on the number of cattle (or other animals) and how it is going to be used. As an example, a farmer with eight cattle and a six-member family wishes to produce gas for cooking and lighting and, if possible, for running a 3hp water pump engine for about an hour every day.

Some of the questions the farmer must ask and guidelines for answering them are:

14

1. How much gas can be expected per day from both eight head
    of cattle and six people?

    Since each cow produces, on the average, 10kg of manure
    per day and 1kg of fresh manure can give .05 cubic meter
    gas, the animals will give 8 X 10kg/animal X .05 cubic
    meter/kg = 4.0 cubic meters gas.

    Each person produces an average of 1kg of waste per day;
    therefore, six people X 1kg/person X .05 cubic meter/kg =
    .30 cubic meter gas.

    The size of the plant would be a 4.3 cubic meter gas
    plant.

2. How much gas does the farmer require for each day?

    Each person requires approximately 0.6 cubic meters gas
    for cooking and lighting. Therefore, 6 X 0.6 = 3.6 cubic
    meters gas.

    An engine requires 0.45 cubic meters gas per hp per hour.
    Therefore, a 3hp engine for one hour is: 3 X 0.45 = 1.35
    cubic meters gas.

    Total gas consumption would be almost 5 cubic meters per
    day--somewhat more than could be produced. Running the
    engine will thus require conserving on lighting and
    cooking (or vice versa), especially in the cool season
    when gas production is low.

3. What will be the volume of the fermentation tank or pit
    needed to handle the mixture of manure and water?

The ratio of manure and water is 1:1.

        8 cattle = 80kg manure + 80kg water = 160kg
        6 people = 6kg waste + 6kg water = 12kg
                          Total input per day = 172kg

Input for six weeks = 172kg X 42 days = 7224kg

1000kg = 1 cubic meter

7224kg = 7.2 cubic meters

Therefore, the minimum capacity of the fermentation well is approximately 7.0 cubic meters--a figure that does not allow for future expansion of the farmer's herd. If the herd does expand and the farmer continues to put all available manure in the tank, the slurry will exit after a shorter digestion period and gas production will be reduced. (The farmer could curtail addition of raw manure and hold it steady at the eight cattle load.) If money is available and there are no digging problems, it is better to put in an oversized than undersized tank.

4. What size and shape of fermentation tank or pit is required?

The shape of the tank is determined by the soil, subsoil, and water table. For this example, we will assume that the earth is not too hard to dig and that the water table is low--even in the rainy season. An appropriate size for a 7.0 cubic meter tank would be a diameter of 1.5 meters. Therefore, the depth required is 4.0 meters.

5. What should the size of the gas cap be?

The metal drum serving as a gas cap covers the fermentation tank and is the most expensive single item in the whole plant. To minimize the size and to keep the price as low as possible, the drum is not built to accommodate a full day's gas production on the assumption that the gas will be used throughout the day and the drum will never be allowed to reach full capacity. The drum is made to hold between 60 and 70 percent of the volume of the total daily gas production.

70% of 4.3 cubic meters = 3-cubic-meter gas cap required

The actual dimensions of the drum may well be determined by the size of the material locally available. A 1.4-meter-diameter drum 1.5 meters tall would be sufficient for this example. See Table 2 for other digester sizes.

HEATING AND INSULATING DIGESTERS

To reach optimum operating temperatures (30-37°C or 85-100°F), some measures must be taken to insulate the digester, especially in high altitudes or cold climates. Straw or shredded tree bark can be used around the outside of the digester to provide insulation. Other forms of heating can also be used such as solar water heaters or the burning of some of the methane produced by the digester to heat water that is circulated through copper coils on the inside of the digester. Solar or gas heating will add to the cost of the digester, but in cold climates it may be necessary. Consult "Further Information Resources" for more information.

MATERIALS (For 3-Cubic-Meter Digester)

. Baked bricks, approximately 3200

. Cement, 25 bags (for foundation and wall covering)

. Sand, 12 cubic meters

. Clay or metal pipe, 20cm diameter, 10 meters

. Copper wire screening (25cm X 25cm)

. Rubber or plastic hose (see page 00)

. Gas outlet pipe, 3cm diameter (see page 00)

. Pipe, 7.5cm diameter, 1.25 meters (gas cap guide)

Gas Plant Type (Model)	Number of Animals	1:1 Water & Dung Per Day (kg)	Volume of Well for 42 Day Digesting (cu m)	Size of Well Diameter & Depth (m)	Size of Gas Cap Diameter & Height (m)	G.I. Sheet for Gas Cap (sq m)	Number of Bricks	Number of Bags Cement (50kg)	Quantity of Sand (cu m)	Gas Produced Per Day (cu m)	Sun Dried Fertilizer Produced Per Day (kg)	Number of People Served by Gas (Cooking, Lighting)
2 cubic meter	4	80	3.5	1.25X3	1.15X1	4.5	2800	22	9	2	4-8	4-5
3 cubic meter	6	120	5	1.5X3.4	1.4X1.25	9	3200	25	12	3	6-12	6-8
4 cubic meter	8	160	7	1.5X4	1.5X1.5	9	4000	28	12	4	8-16	9-11
5 cubic meter	10	200	8.5	1.7X3.5	1.6X1.5	10.5	4000	30	14	5	10-20	12-15
7.5 cubic meter	15	300	13	2X4	1.9X1.5	12.6	5200	32	16	7.5	15-30	15-20
10 cubic meter	20	400	17	2.2X4.3	2.1X1.5	14.3	6400	35	18	10	20-40	20-30

Table 2. Measurements for a Number of Simple Gas Plants

. Pipe, 7cm diameter, 2.5 meters (center guide)

. Mild steel sheeting, .32mm (30 gauge) to 1.63mm (16 gauge), 1.25 meters X 9 meters long

. Mild steel rods, approximately 30 meters (for bracing)

. Waterproof coating (paint, tar, asphalt, etc.), 4 liters (for gas cap

TOOLS

. Welding equipment (gas cap construction, pipe fittings, etc.)

. Shovels

. Metal saw and blades for cutting steel (welding equipment may be used)

. Trowel

V. CONSTRUCTION

PREPARE FOUNDATION AND WALLS

. Dig a pit 1.5 meters in diameter to a depth of 3.4 meters.

. Line the floor and walls of the pit with baked bricks and bound it with lime mortar or clay. Any porousness in the construction is soon blocked with the manure/water mixture. (If a water table is encountered, cover the bricks with cement.)

. Make a ledge or cornice at two-thirds the height (226cm) of the pit from the bottom. The ledge should be about 15cm wide for the gas cap to rest on when it is empty (see Figure 3). This ledge also serves to direct into the gas cap any gas forming near the circumference of the pit and prevents it from escaping between the drum and the pit wall.

. Extend the brickwork 30-40cm above ground level to bring the total depth of the pit to approximately 4 meters.

. Make the input and output piping for the slurry from ordinary 20cm clay drainpipe. Use straight input piping. If the pipe is curved, sticks and stones dropped in by playful children may jam at the bend and cannot be removed without emptying the whole pit. With straight piping, such objects can fall right through or can be pushed out with a piece of bamboo.

. Have one end of the input piping 90cm above ground level and the other end 70cm above the bottom of the pit (see Figure 3).

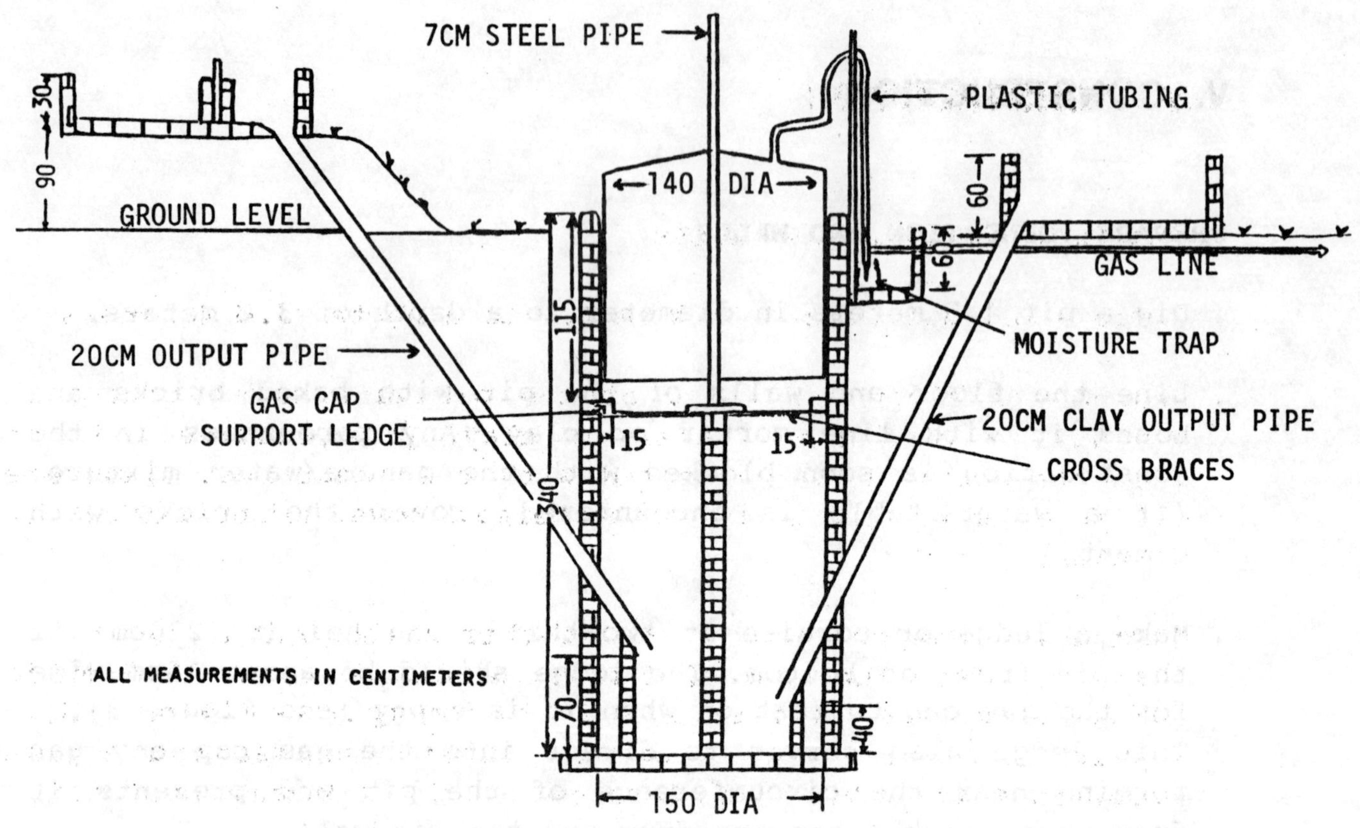

Figure 3. 3-Cubic-Meter Gas Digester

. Have one end of the output piping 40cm above the bottom of
the pit opposite the input pipe and the other end at ground
level.

. Put an iron or wire strainer (copper screening) with 0.5cm
holes at the upper end of the input and the output pipes to
keep out large particles of foreign matter from the pit.

. Construct a center wall that divides the pit into two equal
compartments. Build the wall to a height two-thirds from the
bottom of the digester (226cm). Build the gas cap guide in
the center top of the wall by placing vertically a 7cm X 2.5
meters long piece of metal piping.

. Provide additional support for the pipe by fabricating a
cross brace made from mild steel.

PREPARE THE GAS CAP DRUM

. Form the gas cap drum from mild steel sheeting or galvanized
iron sheeting of any thickness from .327mm (30 gauge) to
1.63mm (16 gauge).

. Make the height of the drum approximately one-third the depth
of the pit (1.25-1.5 meters).

. Make the diameter of the drum 10cm less than that of the pit
(1.4 meters diameter) as shown in Figure 4.

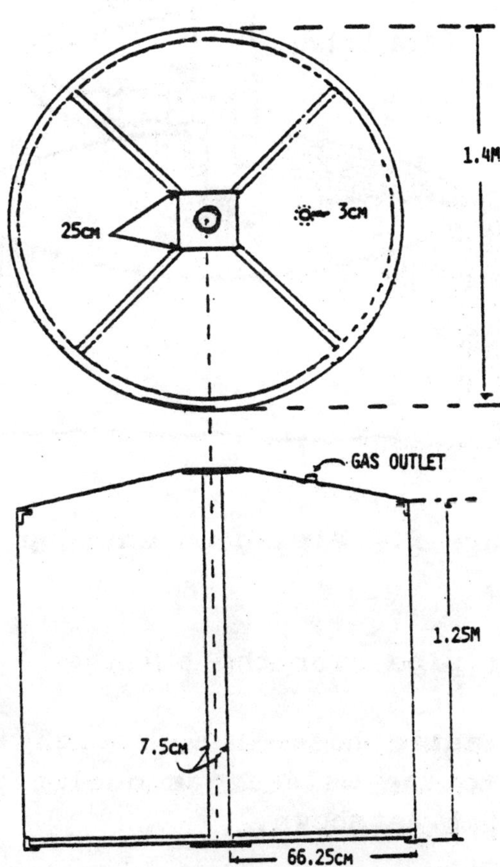

Figure 4. Biogas Plant Gas Cap

. Using a flange, attach a 7.5cm pipe to the inside top center.

. Fix the lower end of the pipe firmly in place with thin, iron
tie rods or angle iron. The cap now looks like a hollow drum
with a pipe, firmly fixed, running through the center.

. Cut a 3cm diameter hole, as shown in Figure 5, in the top of
the gas cap.

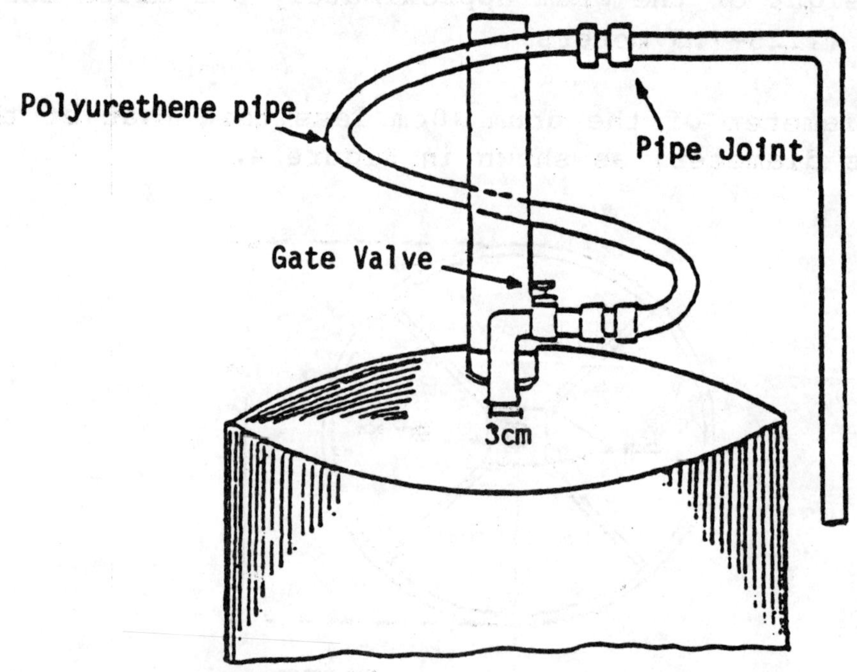

Figure 5. Piping on Gas Cap

. Weld a 3cm diameter pipe over the hole.

. Fix a rubber or plastic hose--long enough to allow the drum
to rise and fall--to the welded gas outlet pipe. A valve may
be fixed at the joint as shown.

. Paint the outside and inside of the drum with a coat of paint
or tar.

. Make sure the drum is airtight. One way to check this is to fill it with water and watch for leaks.

. Turn the gas cap drum so that the outlet pipe is on top and slip the 7.5cm pipe fixed in the gas cap over the 7cm pipe fixed in the center wall of the pit. When empty, the drum will rest on the 15cm ledges built on either side. As gas is produced and the drum empties and fills, it will move up and down the center pole.

. Attach handles to either side of the drum. These don't have to be fancy, but they will prove very helpful for lifting the drum off and for turning the drum.

. Weld a 10cm wide metal strip to each of the tie rod supports in a vertical position. These "teeth" will act as stirrers. By grasping the handles and rotating the drum it is possible to break up troublesome scum that forms on the slurry and tends to harden and prevent the passage of gas.

PREPARE MOISTURE TRAP

. Place a jar of water outside the pit and put into it the end of a downward projection of the gas pipe at least 20cm long. Any moisture condensing in the pipe flows into the jar instead of collecting in the pipe and obstructing the passage of gas. Water then overflows and is lost in the ground. Remember to keep the jar full or the gas will escape. An ordinary tap when opened lets the water escape. Whether using the water jar or tap, do not let the length be greater than 30cm below ground level or it becomes too difficult to reach (see Figure 3 on page 20).

PREPARE THE MIXING AND EFFLUENT TANKS

. Build or improvise a mixing tank to be placed near the out-side opening of the inlet pipe. Likewise, provide a container

at the outlet to catch the effluent. Some provision may also
be made for drying the effluent as the plant goes into full
production.

VI. OPERATION

In order to start up the new digester, it is necessary to have 3 cubic meters (3000kg) of manure. In addition, approximately 15kg of "seeder" is required to get the bacteriological process started. The "seeder" can come from several sources:

. Spent slurry from another gas plant

. Sludge or overflow water from a septic tank

. Horse or pig manure, both rich in bacteria

. A 1:1 mixture of cow manure and water that has been allowed to ferment for two weeks

Put the manure and "seeder" and an equal amount of water into the mixing tank. Stir it into a thick liquid called a slurry. A good slurry is one in which the manure is broken up thoroughly to make a smooth, even mixture having the consistency of thin cream. If the slurry is too thin, the solid matter separates and falls to the bottom instead of remaining in suspension; if it is too thick, the gas cannot rise freely to the surface. In either case the output of gas is less.

When filling the pit for the first time, pour the slurry equally into both halves to balance the pressure on the thin inner wall, or it may collapse.

Mix 60kg fresh manure with 60kg water and add it to the tank every day.

The advantage of this model is that since the daily flow of slurry goes up the first side, where the insoluble matter rises, and down the second, where this matter naturally tends to fall, the outgoing slurry daily draws out with it any sludge found at the bottom. Thus having to clean out the pit becomes a comparatively rare necessity. Sand and gravel may build up on the bottom of the digester and will have to be cleaned from time to time depending on your location.

It can take four to six weeks from the time the digester is fully loaded before enough gas is produced and the gas plant becomes fully operational. The first drumful of gas will probably contain so much carbon dioxide that it will not burn. On the other hand, it may contain methane and air in the right proportion to explode if ignited. DO NOT ATTEMPT TO LIGHT THE FIRST DRUMFUL OF GAS. Empty the gas cap and let the drum fill again.

At this point the gas is safe to use.

OUTPUT AND PRESSURE

The gas cap drum floating on the slurry creates a steady pressure on the gas at all times. This pressure is somewhat lower than that usually associated with other gases that are under pressure but is sufficient for cooking and lighting.

Table 3, on the following page, shows gas consumption by liters/hour.

1	2	3*
Gas cooking	2" diameter burner	280
	4" diameter burner	395
	6" diameter burner	545
Gas lighting	1 mantle lamps	78
	2 mantle lamps	155
	3 mantle lamps	190
Refrigerator	18" X 18" X 12"	78
Incubator	18" X 18" X 18" Flame operated	
Running engines	Converted diesel	350-550 hp/hr

*Liters/hour

Note: These figures will vary slightly depending on the design
       of the appliance used, the methane content of the gas,
       the gas delivery pressure, etc.

Table 3. Application Specification for Gas Consumption

VII. VARIOUS APPLICATIONS OF BIOGAS AND DIGESTER BY-PRODUCTS

ENGINES

Internal Combustion

Any internal combustion engine* can be adapted to use methane. For gasoline engines, drill a hole in the carbuerator just near the choke and introduce a 5mm diameter tube connected to the gas supply through a control valve. The engine may be started on gasoline then switched over to methane while running, or vice-versa. For smooth running of the engine, the gas flow should be steady. For stationary engines this is done by counterbalancing the gas cap. (Refer to Table 3 on page 17 for gas consumption.)

Diesel

Diesel engines are run by connecting the gas to the air intake and closing the diesel oil feed. A spark plug will have to be placed where the injector normally is and arrangement made for electricity and spark timing. Modifications will vary with the make of the engine. One suggestion is to adapt the full-pump mechanism for timing the spark.

---

*Some authorities recommend that when running the internal combustion engines, the gas be first purified. This is done by bubbling it through lime water, to remove carbon dioxide, and through iron filings, to remove hydrogen sulphide.

FERTILIZER

The sludge product of anaerobic decomposition produces a better fertilizer and soil conditioner than either composted or fresh manure. The liquid effluent contains many elements essential to plant life: nitrogen, phosphorous, potassium, plus small amounts of metallic salts indispensible for plant growth.

Methods of applying this fertilizer are numerous and conflicting. The effluent can be applied to crops as either a diluted liquid or in a dried form. Remember that although 90-93% of toxic pathogens found in raw human manure are killed by anaerobic decomposition, there is still a danger of soil contamination with its use. The effluent should be composted before use if the slurry contains a high proportion of human waste. However, when all factors are considered, the effluent is much safer than raw sewage, poses less of a health problem, and is a better fertilizer.

The continued use of the effluent in one area tends to make soils acidic unless it is duluted with water (3 parts water to 1 part effluent is considered a safe mixture). A little dolomite or crushed limestone added to the effluent containers at regular intervals will cut down on acidity. Unfortunately, limestone tends to evaporate ammonia; so it is generally best to keep close watch over the amount of effluent provided to crops until the reaction of the soil and crops is certain.

IMPROVISED STOVE

Because gas pressure is low, it will be necessary to modify existing equipment or build special burners for cooking and heating. A pressure stove burner will work satisfactorily only after certain modifications are made to the burner. The needle-thin jet should be enlarged to 1.5mm. To make a burner out of 1.5cm water pipe, choke the pipe with a metal disc having a center hole with a diameter of 1.5 to 2mm. An efficient burner is a tin can, filled with stones for balance, having six 1.5mm holes in the top. The gas enters through a pipe choked to a 2mm orifice. Or fill a chula or Lorena stove with stones and insert a pipe choked to a 2mm orifice.

If possible, it is best to use a burner with an adjustable air inlet control. The addition or subtraction of air to the gas creates a hotter flame with better use of available gas.

LIGHTING

Methane gives a soft, white light when burned with an incandescent mantle. It is not quite as bright and glaring as a kerosene lantern. Lamps of various tyles and sizes are manufactured in India specifically for use with methane. Each mantle burns about as bright as a 40-watt electric bulb.

Some biogas appliances manufactured by an Indian firm are:

- Indoor hanging lamp
- Indoor suspension lamp
- Outdoor hanging lamp
- Indoor table lamp

- Stoves and burners
- Bottle syphons and pressure gauges

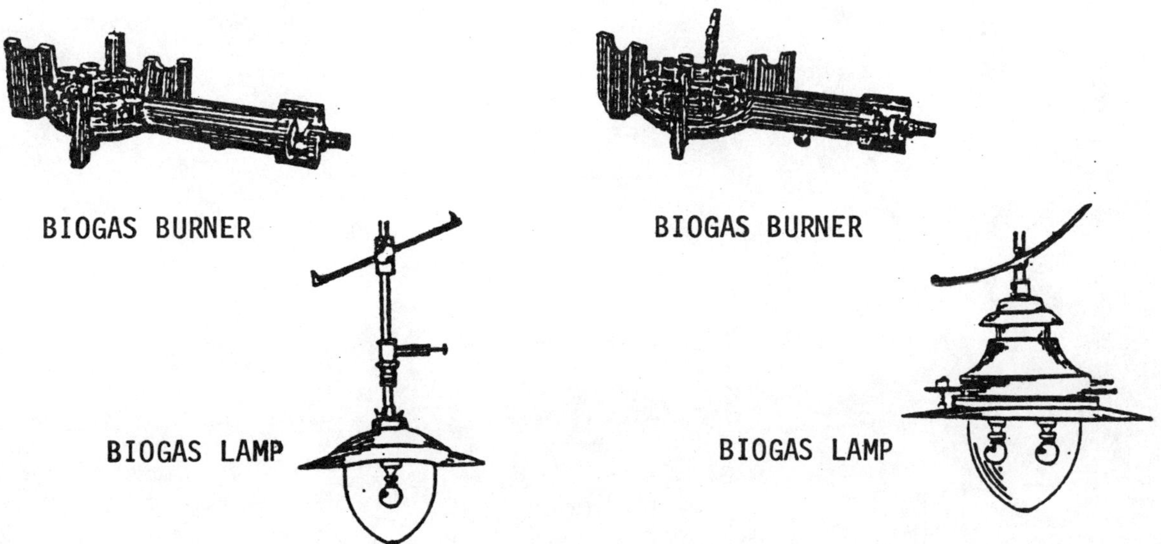

Bengal Scientific & Technical Works (P) Ltd.
20/3 Aswani Dutt Road, Calcutta 29

VIII. MAINTENANCE

A digester of this type is virtually maintenance free and has a life of approximately 25 years. As long as cow or other animal manure is used, there should be no problems. Vegetable matter can also be used for methane production but the process is much more complex. Introduction of vegetable matter in the digester is <u>not recommended</u>.

A trouble-shooting guide is listed below for possible problems that may be encountered.

POSSIBLE TROUBLES

<u>Defect</u>	<u>May be caused by</u>	<u>Remedy</u>
No gas. Drum won't rise.	a) No bacteria	Add some bacteria (seeder)
	b) Lack of time	Patience! Without bacteria, it may take four or five weeks.
	c) Slurry too cold	Use warm water. Cover plant with plastic tent or use heating coil.
	d) Insufficient input	Add right amount of slurry daily.
	e) Leak in drum or pipe	Check seams, joints, and taps with soapy water.
	f) Hard scum on slurry blocking gas.	Remove drum; clean slurry surface. With sliding-drum plants, turn drum slightly to break crust.

No gas at stove; plenty in drum.	a) Gas pipe blocked by condensed water	Open escape cock.
	b) Insufficient pressure	Increase weight on drum
	c) Gas inlet blocked by scum	Remove drum and clean inlet. Close all gas-taps. Fill gas line with water; apply pressure through moisture escape. Drain water.
Gas won't burn.	a) Wrong kind is being formed.	Slurry too thick or too thin. Measure accurately. Have patience.
	b) Air mixture	Check burner gas jet to make sure it is at least 1.5mm.
Flame soon dies.	a) Insufficient	Increase weight on drum.
	b) Water in line	Check moisture escape jar. Drain gas line.
Flame begins far	a) Pressure too high	Remove weights from drum. Counterbalance.
	b) Air mixture	Choke gas inlet at stove to 2mm (thickness of 1" long nail).

IX. TEST GAS LINES FOR LEAKS

Checking for gas leaks is done by closing all gas taps, including the main gas tap beside the gas holder, except for one.

Then to the open tap, a clear plastic pipe about a meter long is attached, and a "U" is formed. The lower half of the "U" is filled with water.

Using a pipe attached to a second tap, pressure is applied until the water in the two legs of the "U" is different by 15cm. The second tap is then closed. The "U" is now what is called a "manometer."

If the water levels out when the second tap is closed, a leak is indicated and can be sought out by putting soapy water over possible leaks, such as joints, in the pipework.

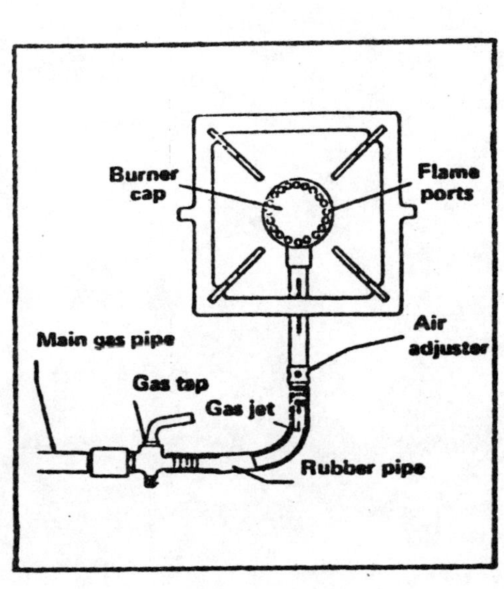

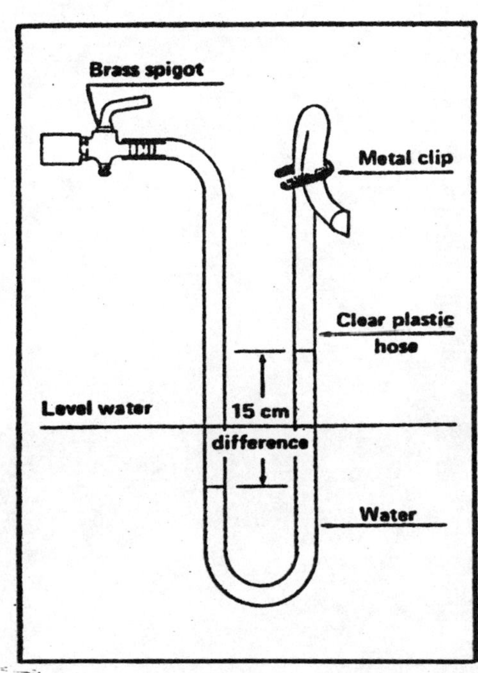

X. DICTIONARY OF TERMS

AEROBIC--Decomposing with oxygen.

ANAEROBIC--Decomposing without oxygen.

BY-PRODUCT--Something produced from something else.

CARBON DIOXIDE--A colorless, odorless, incombustible gas (CO_2) formed during organic decomposition.

DECOMPOSE--To rot, to disintegrate, to breakdown into component parts.

DIA (DIAMETER)--A straight line passing completely through the center of a circle.

DIGESTER--A cylindrical vessel in which substances are decomposed.

EFFLUENT--The outflow from the biogas storage tank.

FERMENT--To cause to become agitated or turbulent.

HP (HORSEPOWER)--Unit of power equal to 747.7 watts.

INSOLUBLE--Incapable of being dissolved.

LEACHED--Dissolved and washed out by a percolating liquid.

MANTLE--A sheath of threads that brightly illuminates when heated by gas.

METHANE--An odorless, colorless, flammable gas (CH_4) used as a fuel.

NITRATES--Fertilizers consisting of sodium and potassium nitrates.

NITROGEN--A colorless and odorless gas (N_2) in fertilizers.

ORGANIC WASTES--Waste from living organisms or vegetable matter.

SCUM--A filmy layer of waste matter that forms on top of liquid.

SEEDER--Bacteria used to start the fermentation process.

SEPTIC TANK--A sewage disposal tank in which a continuous flow of waste material is decomposed by anaerobic bacteria.

SLUDGE--A thick liquid composed of 1:1:1 mixture of manure, seeder, and water.

SUPERNATANT--Floating on the surface.

TOXIC PATHOGENS--Harmful or deadly agents that cause serious disease or death.

XL. CONVERSION TABLES

UNITS OF LENGTH

1 Mile	= 1760 Yards	= 5280 Feet
1 Kilometer	= 1000 Meters	= 0.6214 Mile
1 Mile	= 1.607 Kilometers	
1 Foot	= 0.3048 Meter	
1 Meter	= 3.2808 Feet	= 39.37 Inches
1 Inch	= 2.54 Centimeters	
1 Centimeter	= 0.3937 Inches	

UNITS OF AREA

1 Square Mile	= 640 Acres	= 2.5899 Square Kilometers
1 Square Kilometer	= 1,000,000 Square Meters	= 0.3861 Square Mile
1 Acre	= 43,560 Square Feet	
1 Square Foot	= 144 Square Inches	= 0.0929 Square Meter
1 Square Inch	= 6.452 Square Centimeters	
1 Square Meter	= 10.764 Square Feet	
1 Square Centimeter	= 0.155 Square Inch	

UNITS OF VOLUME

1.0 Cubic Foot	= 1728 Cubic Inches	= 7.48 US Gallons
1.0 British Imperial Gallon	= 1.2 US Gallons	
1.0 Cubic Meter	= 35.314 Cubic Feet	= 264.2 US Gallons
1.0 Liter	= 1000 Cubic Centimeters	= 0.2642 US Gallons

```
1.0 Metric Ton        = 1000 Kilograms        = 2204.6 Pounds
1.0 Kilogram          = 1000 Grams            = 2.2046 Pounds
1.0 Short Ton         = 2000 Pounds
```

UNITS OF PRESSURE

```
1.0 Pound per square inch              = 144 Pound per square foot
1.0 Pound per square inch              = 27.7 Inches of water*
1.0 Pound per square inch              = 2.31 Feet of water*
1.0 Pound per square inch              = 2.042 Inches of mercury*
1.0 Atmosphere                         = 14.7 Pounds per square inch (PSI)
1.0 Atmosphere                         = 33.95 Feet of water*
1.0 Foot of water  =  0.433 PSI        = 62.355 Pounds per square foot
1.0 Kilogram per square centimeter     = 14.223 Pounds per square inch
1.0 Pound per square inch              = 0.0703 Kilogram per square
                                           centimeter
```

UNITS OF POWER

```
1.0 Horsepower (English)          = 746 Watt   = 0.746 Kilowatt (KW)
1.0 Horsepower (English)          = 550 Foot pounds per second
1.0 Horsepower (English)          = 33,000 Foot pounds per minute
1.0 Kilowatt (KW)   = 1000 Watt   = 1.34 Horsepoer (HP) English
1.0 Horsepower (English)          = 1.0139 Metric horsepower
                                      (cheval-vapeur)

1.0 Metric horsepower             = 75 Meter X Kilogram/Second
1.0 Metric horsepower             = 0.736 Kilowatt   = 736 Watt
```

---

*At 62 degrees Fahrenheit (16.6 degrees Celsius).

XII. FURTHER INFORMATION RESOURCES

A LISTING OF RECOMMENDED RESOURCE MATERIALS

Biogas Plant: Designs With Specifications. Ram Box Singh, Gobar
    Gas Research Statin Ajit Mal Etawah (V.P.) India. The
    main part of this book is taken up with very detailed
    technical drawings of 20 different models of methane
    digesters for various size operatins and different cli-
    mates. Also has designs for gas burners, lamps, and a
    carburator. No real written instructions, but would be
    very useful if used in conjunction with a more general
    manual.

Biogas Plant: Generating Methane from Organic Wastes. Ram Bux
    Singh, Gobar Gas Research Station, Ajitmal Etawah (V.P.)
    India, 1974. The most comprehensive work on biogas. Gives
    the background of the subject, an extensive treatment of
    just how a biogas plant works, factors to consider in
    designing a plant and several designs, and instructions
    for building a plant and using the products. Profusely
    illustrated, this is considered by some as the "bible" of
    biogas.

Fuel Gas From Cow Dung. Bertrand R. Saubolle, S. J., Sahayog;
    Prakashan Tripureshwas, Kathmandu, April 1976, 26 pp.
    Fairly detailed manual for obtaining and using methane
    from cow manure. Includes a trouble-shooting section and
    specification charts for different size digesters. Writ-
    ten in straight forward, nontechnical language. Potential
    quite useful. Available from VITA.

Small-Scale Biogas Plants. Nigel Florida; Bardoli, India.
    Highly detailed manual. Gives step-by-step instructions

for building and operating a methane digester. Includes modifications needed to cope with a variety of conditions and a detailed analysis of digested slurry and of the produced biogas. Also has a chapter on current state-of-the-art in India. Available from VITA.

USEFUL INFORMATION FOR METHANE DIGESTER DESIGNS

Andrews, John F. Start-Up and Recovery of Anaerobic Digestion, 8 pp. Clemson University. Available from VITA.

"Biogas Plant: Generating Methane from Organic Wastes." Compost Science. January-February 1972, pp. 20-25. Available from VITA.

Biogas Stove and Lamp: Efficient Gas Appliances, Examples of Plant Designs, Examples of Biogas Plants, Construction Notes. 4 pp. including illustrations. Available from VITA.

"Building a Biogas Plant." Compost Science. March-April 1972. pp. 12-16. Available from VITA.

Finlay, John H. Operation and Maintenance of Gobar Gas Plants, April 1976, 22 pp. with 3 diagrams. Nepal. Available from VITA.

Gobar Gas Plant, 4 pp. Appropriate Technology Development Association, PO Box 311, Gandhi Bhawan, Lucknow 226001, UP, India.

Gobar Gas Plants, 8 pp. with 4 diagrams. Indian Agricultural Research Institute. Available from VITA.

Gotaas, Harold B. "Manure and Night-Soil Digesters for Methane Recovery on Farms and in Villages. Composting: Sanitary Disposal and Reclamation of Organic Wastes. 1956, chapter 9, pp. 171-199. University of California/Berkeley, World Health Organization. Available from VITA.

Grout, A. Roger. Methane Gas Generation from Manure, 3 pp. Pennsylvania State University. Available from VITA.

Hansen, Kjell. A Generator for Producing Fuel Gas from Manure, 4pp. Available from VITA.

Hill, Peter. Notes on a Methane Gas Generator & Water Tank Construction, June 1974, 9 pp. Belau Modekngei School. Available from VITA.

Information on Cow Dung Gas: A Manure Plant for Villages, 5 pp. Indian Agricultural Research Institute, Division of Soil Science and Agricultural Chemistry, Pusa, New Delhi, India.

Klein, S.A. "Methane Gas--An Overlooked Energy Source." Organic Gardening and Farming, June 1972, pp. 98-101. Rodale Press, Inc., 33 East Mine Street, Emmaus, Pennsylvania 18049 USA.

Oberst, George L. Cold-Region Experiments with Anaerobic Digestion for Small Farms and Homesteads. Biofuels, Box 609, Noxon, Montana 59853 USA.

The Pennsylvania State University Digester-Methane Generator, 2 pp. Available from VITA.

Shifflet, Douglas. Methane Gas Generator, 1966. Available from VITA.

Vani, Seva. "Mobile Gobar Gas Plant," Journal of CARITAS India, January-February 1976, 2 pp. Available from VITA.

APPENDIX I

DECISION MAKING WORKSHEET

If you are using this as a guideline for using a biogas plant in a development effort, collect as much information as possible and if you need assistance with the project, write VITA. A report on your experiences and the uses of this manual will help VITA both improve the book and aid other similar efforts.

Volunteers in Technical Assistance
3706 Rhode Island Avenue
Mt. Rainier, Maryland 20822 USA

CURRENT USE AND AVAILABILITY

. Note current domestic and agricultural practices that might benefit from a biogas plant: improved fertilizer, increased fuel supply, sanitary treatment of human and animal wastes, etc.

. Have biogas plant technologies been introduced previously? If so, with what results?

. Have biogas plant technologies been introduced in nearby areas? If so, with what results?

. What changes in traditional thinking or practices might lead to increased acceptance of biogas plants? Are such changes too great to attempt now?

. Under what conditions would it be useful to introduce biogas plant technology for demonstration purposes?

• If biogas plants are feasible for local manufacture, would they be used? Assuming no funding, could local people afford them? Are there ways to make the biogas plant technologies pay for themselves?

• Could this technology provide a basis for a small business enterprise?

NEEDS AND RESOURCES

• What are the characteristics of the problem? How is the problem identified? Who sees it as a problem?

• Has any local person, particularly someone in a position of authority, expressed the need or showed interest in biogas plant technology? If so, can someone be found to help the technology introduction process? Are there local officials who could be involved and tapped as resources?

• Based on descriptions of current practices and upon this manual's information, identify needs that biogas plant technologies appear able to meet.

• Do you have enough animals to supply necessary amount of manure needed daily?

• Are materials and tools available locally for construction of biogas plants?

• What would be the main use of the methane produced by the biogas plant? For example, heating, lighting, cooking, etc.

• Would you be able to use all of the effluent fertilizer or would you have more than you need? Would you be able to sell the surplus?

• Do a cost estimate of the labor, parts, and materials needed.

. What kinds of skills are available locally to assist with construction and maintenance? How much skill is necessary for construction and maintenance? Do you need to train people in the construction techniques? Can you meet the following needs?

-- Some aspects of the project require someone with experience in metal-working and/or welding.

-- Estimated labor time for full-time workers is:

. Skilled labor - 8 hours
. Unskilled labor - 80 hours
. Welding - 12 hours

. How much time do you have? When will the project begin? How long will it take?

. How will you arrange to spread knowledge and use of the technology?

FINAL DECISION

. How was the final decision reached to go ahead--or not to go ahead--with this technology?

APPENDIX II

RECORD KEEPING WORKSHEET

CONSTRUCTION

Photographs of the construction process, as well as the fin-
ished result, are helpful. They add interest and detail that
might be overlooked in the narrative.

A report on the construction process should include very specific information. This kind of detail can often be monitored
most easily in charts (such as the one below).

<u>CONSTRUCTION</u>

Labor Account

Hours Worked

Name	Job	M	T	W	T	F	S	S	Total	Rate?	Pay?
1											
2											
3											
4											
5									Totals		

Materials Account

	Item	Cost Per Item	# Items	Total Costs
1				
2				
3				
4				
5				
			Total Costs	

Some other things to record include:

. Specification of materials used in construction.

. Adaptations or changes made in design to fit local conditions.

. Equipment costs.

. Time spent in construction--include volunteer time as well as paid labor, full- and/or part-time.

. Problems--labor shortage, work stoppage, training difficulties, materials shortage, terrain, transport.

OPERATION

Keep log of operations for at least the first six weeks, then periodically for several days every few months. This log will vary with the technology, but should include full requirements, outputs, duration of operation, training of operators, etc. Include special problems that may come up--a damper that won't close, gear that won't catch, procedures that don't seem to make sense to workers, etc.

MAINTENANCE

Maintenance records enable keeping track of where breakdowns occur most frequently and may suggest areas for improvement or strengthening weakness in the design. Furthermore, these records will give a good idea of how well the project is working out by accurately recording how much of the time it is working and how often it breaks down. Routine maintenance records should be kept for a minimum of six months to one year after the project goes into operation.

<u>MAINTENANCE</u>

Labor Account

Name	Hours & Date	Repair Done	Also down time Rate?	Pay?
1				
2				
3				
4				
5				
Totals (by week or month)				

Materials Account

Item	Cost	Reason Replaced	Date	Comments
1				
2				
3				
4				
5				
Totals (by week or month)				

SPECIAL COSTS

This category includes damage caused by weather, natural disasters, vandalism, etc. Pattern the records after the routine maintenance records. Describe for each separate incident:

. Cause and extent of damage.
. Labor costs of repair (like maintenance account).
. Material costs of repair (like maintenance account).
. Measures taken to prevent recurrence.

ABOUT VITA

Volunteers in Technical Assistance (VITA) is a private, non-profit, international development organization. It makes available to individuals and groups in developing countries a variety of information and technical resources aimed at fostering self-sufficiency--needs assessment and program development support; by-mail and on-site consulting services; information systems training.

VITA promotes the use of appropriate small-scale technologies, especially in the area of renewable energy. VITA's extensive documentation center and worldwide roster of volunteer technical experts enable it to respond to thousands of technical inquiries each year. It also publishes a quarterly newsletter and a variety of technical manuals and bulletins.

VITA's documentation center is the storehouse for over 40,000 documents related almost exclusively to small- and medium-scall technologies in subjects from agriculture to wind power. This wealth of information has been gathered for almost 20 years as VITA has worked to answer inquiries for technical information from people in the developing world. Many of the documents contained in the Center were developed by VITA's network of technical experts in response to specific inquiries; much of the information is not available elsewhere. For this reason, VITA wishes to make this information available to the public.

For more information, contact VITA at 3706 Rhode Island Avenue, Mount Rainier, Maryland 20822 U.S.A.